GET MOBILE

The Essential Guide To Building Your Own Mobile Real Estate Office

by Brian Cross

Note for Librarians: A cataloguing record for this book is available from Library
and Archives Canada at www.collectionscanada.ca/amicus/index-e.html

ISBN 1-4251-0383-9

Printed on paper with minimum 30% recycled fibre.
Trafford's print shop runs on "green energy" from solar, wind and other environmentally-friendly
power sources.

Offices in Canada, USA, Ireland and UK

Book sales for North America and international:
Trafford Publishing, 6E–2333 Government St.,
Victoria, BC V8T 4P4 CANADA
phone 250 383 6864 (toll-free 1 888 232 4444)
fax 250 383 6804; email to orders@trafford.com
Book sales in Europe:
Trafford Publishing (UK) Limited, 9 Park End Street, 2nd Floor
Oxford, UK OX1 1HH UNITED KINGDOM
phone +44 (0)1865 722 113 (local rate 0845 230 9601)
facsimile +44 (0)1865 722 868; info.uk@trafford.com
Order online at:
trafford.com/06-2140

10 9 8 7 6 5 4 3 2 1

Dedication

My mother whose love, encouragement, insight and mentorship have inspired me beyond words.

My entire family and extended family including the Krones, the Conleys, the Kennedys, the Shurkens, the Houges, the Royals, the Germanys and of course, the entire Cross family. All of you mean the world to me.

My long time friend Todd Bogert who is The King of All Computer Geeks. Thanks for teaching me what an ID10-T error is.

My father, my uncle Mike and my brother Dave who are no longer with us, and will always be remembered with a loving heart.

Table of contents

Disclaimer

*This book contains suggestions for products and services. These suggestions in no way are an endorsement or guarantee for any of the products or services. They are representations of my personal experience while working in the mobile office environment. Always do your own research and choose the best provider or product for you.

You may already have some of the tools described in this manual. If you are comfortable with a particular product or service you are already using, continue using it! "The best technology is the one you use!" Make your mobile office your own.

1. Introduction

The scary and often times confusing aspect of technology is that there is so much out there, and knowing what you need can be frustrating, not to mention time consuming. The goal of this book is to offer you my personal experience in building a successful mobile real estate office with minimal trial and error on your part.

Mobile office technology is perfectly suited to a handful of industries. Real estate happens to be one of them. Instant access to information in the field will become a buyer expectation and could be the difference of writing a contract or losing a client. Selling real estate necessitates the need for mobile technology because the houses are "in the field." An accountant for example might not have the same need for mobile technology.
If you have had experience driving buyers around, I can almost guarantee you have heard the phrase "what about that house?" This is always

accompanied by the client pointing to house that you have no information on. It's a truly sinking feeling.

Enter the laptop. At first my laptop rode shotgun in the passenger's seat and was moving around with every turn and stop, not to mention it was hard to reach. I remember hoping my precious laptop wouldn't fly off the seat as a result of a quick stop or turn. "There must be a better way!"

There are many technologies out there for real estate agents. Learn from my experience. You don't need them all. *Marry the technology to your business, not your business to technology.*

The following pages are the result of research and trial. As with working with computers, there are always more ways than one to get the job done. The products, services and methods contained in these pages are suggestions for an individual who wants more than a PDA to be mobile.

2. Before You Get Started

Building a mobile office does not require a high level of technical computer skills. There are however basic computer skills you must master before using your mobile office with confidence.

Never practice in front of your clients! The client is looking to you to have the skill sets necessary to guide them through the transaction process. Practice makes perfect!

Also, it helps if you have already done a few transactions before starting out with a mobile office. This would include presenting an offer on behalf of your client and seeing it through closing.

3. Basic Computer Skills Needed

*Setting up a computer out of the box (hooking up printer, keyboard etc..)

*Copying and pasting files

*Creating new folders

*Downloading and installing programs from the Internet or a CD

*Using Zip Form® or similar software

There are many resources for learning

*Your local Real Estate board
*A local community college
*Your office or brokerage
*Mentors
*Online education

4. Everything You Need

Before we get into specifics, let's first talk about the basic elements. More about each element will be described in the pages to follow.

1. A reliable, comfortable vehicle

2. A laptop, tablet, or ultra portable PC

3. A mounting device for your computer

4. A mobile Internet connection

5. A printer, preferably one that also makes copies

6. Various cords & power inverters

7. Various software

5. How much will it cost?

At first, the concept of a mobile office sounds like an expensive endeavor. For around $2600 you can get your mobile office up and running! Do the math and figure the time it will take to recoup your costs. I bet it is surprisingly small.

The following is a detailed list of the elements of your mobile real estate office along with cost estimates. To put an exact figure on making a mobile office would be difficult, since there could be many variations to your particular configuration.

Item Description	Cost
Vehicle	Up to you
Tablet PC or laptop	$500-$2000
Mounting device for computer	$300-$400
Wireless Internet card	Around $100 for the card + around $60 per month for service
Cords & power inverters	$200-$300
E-fax service	$10-$20 per month depending on plan
Multi-functional printer	$250-$600
GPS	$150-$2000
Installation	$0-$350
Estimated startup cost including computer but not including vehicle	**$2600**

6. Selecting a vehicle

For your first mobile office, you can go a number of routes. You can certainly use your existing car. If this isn't possible, you can purchase a car to be used solely as your mobile office.

There are a few things to consider when looking for a suitable vehicle. While almost any vehicle can be configured into a mobile office, there are a few key points to consider when choosing your vehicle:

- Comfort for your client, ease of getting in and out
- Gas mileage
- Interior space
- Reliability
- Built in power outlets (not required but helpful)

Minivan- This is probably the most sensible and functional solution. There is sufficient room, and they are fairly easy to get in and out of. Try looking for a van that's not too high off the ground. (buyers get tired looking at multiple houses.) Try a van with remote electric sliding doors on each side. It's very impressive to open doors for your clients. I am presently using a 2006 Chevy Uplander LT.

Full size van- If you have the place to park it, consider getting a full size conversion van. Leather captain chairs, built in monitors and plenty of room indicate that you are concerned with your client's comfort.

Passenger van- If you are handy or have a friend who is, a full size passenger van can give you the space and feel of a real office complete with desk space and office chairs. Imagine a cherry wood wrap-around desk with comfortable seating for you & your clients. Your imagination is the limit.

Sedan- If you already have a sedan and want to use it as your mobile office, that's ok. However, with this configuration, you will need to purchase a smaller, ultra-portable laptop. (more information on this in the Hardware chapter)

SUV- SUV's offer more interior space than sedans. Be careful not to get one too high off the ground. When getting in and out of the car all day, a lower vehicle will be much more comfortable for your clients.

7. Hardware

Tablet PC

A Tablet PC works in almost the exact same manner as a regular laptop. The main difference being, you can swing the screen around 180 degrees, lay it flat and write directly on the screen using a stylus. Tablet PCs are also usually slimmer and smaller than traditional laptops. Tablet PC's are more expensive, but if you are going to make the investment, it's well worth it. More about the advantages of signing on the screen later.

A few great Tablet PC manufacturers are HP, Toshiba, Panasonic, and Fujitsu. I presently use a Toshiba Protégé and have been very happy with its performance.

There is also what is called "rugged" Tablet PCs. These are more expensive, but offer features such as glare resistant screens and higher temperature ratings for your car. A rugged computer is not mandatory, but will hold up to more punishment over the years.

*Notice- You don't necessarily have to have a Tablet PC if you've already purchased a laptop. A Tablet PC will offer more features in terms of digital signing and paperless transactions.

Air card

Forget about "hot spots!" Hot spots are areas in coffee shops and airports that allow you to get online. (usually for a fee) Technology now exists that will allow you to be mobile from anywhere such as your client's home, parked on the street, or in front of a home your client is interested in. The air card is beginning to gain popularity. Your computer will have the ability to be online from almost anywhere. This would include coverage in almost any major city.

Service may not be available in all areas, so check with your local providers. Some companies which offer this service are Sprint, Cingular, and Verizon.

These cards fit directly into your laptop's PC card slot. Installing the software is easy, and the service is for the most part reliable. You then simply "dial-up" on the network using the air card software. You now have instant access to anything from anywhere!

Mounting device for your computer

Your Tablet PC or laptop must be securely mounted at all times in your mobile real estate office. A laptop sliding around is unprofessional and is dangerous to your clients and can damage the hardware. For most mobile office configurations, a Jotto Desk® is ideal. Jotto Desk® is a company that manufactures mounting equipment for mobile computers. They can be found at www.jottodesk.com. Which mounting product you choose depends on the type of car or van you choose. Once your "desk" is mounted, it can be swiveled in any direction.

Multi-functional printer

Why a multi function all in one printer? Because any office should have the ability to print, copy and scan documents. The printing and coping functions are particularly useful in the field. You want to be able to complete your part of the transaction in your mobile office.

There are small mobile printers on the market. I have found a full-size printer to be much more efficient. Some smaller printers hold limited amounts of ink, and running out of ink in the field is the last thing you want.

Lexmark makes a great combination unit and it's inexpensive at only around $250. The printer can be placed in the trunk or it can sit behind the 3rd row of seats, if you have a van. (see diagrams on pages 35-36) If you feel you need to mount your printer, consult a professional audio/video installation company. I have found my Lexmark fits neatly behind the 3rd row seats without any mounting device.

*Caution- If you live in a hot weather environment, don't leave your printer cartridges in your machine for a long period of time. The ink will evaporate, and ink is expensive!

GPS

Global Positioning Satellite (GPS) is particularly useful in real estate. There are two routes to go. You can either purchase a stand-alone GPS unit which mounts on your dash, or purchase the software and GPS receiver for your laptop or Tablet PC. Purchasing a unit that will work with your existing PC is much less expensive, and your computer's screen is obviously larger than a stand-alone unit. Let's say you are looking at multiple houses with a client. You can input all of the addresses and the GPS will tell you the shortest routes to take.

Cords

The only specialty cord you may need is a long USB cable to run from your printer in the back, up to your laptop or hub in the front. You may also connect your computer to your printer using either Blue Tooth or Wi-Fi methods if you have a wireless compatible printer. I purchased two USB extension cords to make the distance from the back of my van to the front where the computer is located. Long USB cords can be purchased at any electronics store.

Remember to have all of your cords out of sight as much as possible. A professional installer can run your USB cord underneath carpeting and molding trim to make sure it's out of sight and out of mind. The standard power cords that came with your computer and printer should work fine.

Using a full-size computer

If you are building actual desk space in a large van, you can use a full size desktop PC. Be careful because a desktop computer will require more power than a Tablet PC or a laptop. There are configurations with a dedicated battery for the computer itself. (see the chapter on Powering Up)

The computer unit itself can be mounted under the desk to maximize space, and a flat screen monitor can be mounted to the desk. You will also need a PCMCIA card reader with a USB cord running into your computer in order to be online and mobile. If you're not sure what this is, consult with a local computer electronics store.

8. Powering Up

With all of this hardware, power is going to be an issue. When it comes to powering your laptop, printer or anything else, it is best to have a professional installer put in the wiring. Many cars already come with 24 volt plugs already installed in the car. If not, you will need a power inverter.

Power Inverters

What is a power inverter? This is a device that will allow you to convert the DC power from your car's battery to the AC power needed with laptops and other appliances. Good converters can be purchased at an auto supply or car audio/video store. The inverter can be mounted underneath a seat, mounted under the dash, or anywhere else it is out of sight.
Consult a local car audio/video installation shop to get it properly wired and mounted. I have found that when it comes to power inverters, you get what you pay for. Do yourself a favor & pay

for a good one. I would recommend getting at least 900 watt unit. Try a local auto supply store.

Depending on how you set up your mobile office, you may need up to two inverters; one up front for the laptop and one in the back for your printer. If you are setting up your printer near the computer, you may only need one power inverter with two plugs.

The power inverter will be required to be hooked up to your vehicle battery either through wiring under the dash or through a cigarette lighter. Your placement of your printer/fax/copier and power inverter will vary depending on your type of car. Most good power inverters will come with bare wire ends, jumper cable clip type ends, and a cigarette lighter adapter.

<u>If you must run a power cord under the carpet from the rear to where the battery is, it's best to hire a professional.</u> One car battery will be sufficient to run your printer and laptop. Many cars come

with power terminals under the dash. Again, it is best to consult a qualified installer or mechanic for this portion. It's also important to secure your inverter to the vehicle where it is out of sight. Once your inverter is hooked to the power source and secured, plug in your laptop and printer and you're ready to go.

You may be thinking of using two batteries for more power. Be careful. In order for a car battery to recharge, it must be hooked up your cars alternator. Hooking up two batteries to an ordinary alternator can damage it. <u>If you must have two power sources, talk to a professional mechanic about the configuration of your electrical system.</u>

9. E-Faxing

E-Faxing is one of the biggest secrets and one of the biggest keys to any mobile office. The cost is minimal at around $10 per month, and the time it will save you is invaluable. There are two companies I would recommend;
www.efax.com
www.protus.com

First, you will need Adobe Acrobat Reader® to view PDF documents. You can get a free version at www.adobe.com.

Here's how e-faxing works:

SIGNING UP
You can either choose a local number or an 800 number for your new fax number. I use an 800 number & my clients find it quite convenient.
INBOUND FAXING
When you sign up, the company will issue you a fax number. When someone faxes you any document, it will come directly to your email inbox as a PDF

Adobe Acrobat® file. This makes organization easy because you can save all of your faxes on your hard drive under your client's name.

OUTBOUND FAXING
When you sign up for one of these services, the company will give you their email address for faxing. All you do is send an email with the attached file you want to fax & the other party is faxed the document right away. It will send the pages in the order that they are attached in your email. A good idea would be to make your fax coversheet, save it, then attach it first, and then attach the other documents.

10. "Wrapping" Your Vehicle

Vehicle wrapping is one of the most effective forms of advertising, especially in real estate. It demands attention, and you should be proud that you have a mobile real estate office. It will not harm your car & can be removed fairly easily.

Wrapping your vehicle is literally wrapping your car like a Christmas present, but with logos, high impact graphics and contact information. You can also do a partial wrap depending on how much you want to spend. An example of a partial wrap may be just placing graphics on the windows. Wrapping your car is not totally necessary for your mobile office, but it is a great attention grabber. What's more, the value of the mobile office idea lies in the vehicle itself. I can't think of a better application for wrapping a vehicle. You will have a moving billboard!

The wrap itself is actually made of vinyl. It prints out in huge sheets. The sheets have a peel-off backing like a sticker. The wrap is professionally cut and applied.

You will need to find a good wrapping company with an excellent reputation. Finding these companies is easy. Just look in the Yellow Pages under categories like auto customizing, auto graphic design, or signs and banners. It's also a good idea to check with your local Better Business Bureau's website to verify the company is in good standing. The company you choose must be able to help you with the design, production and installation.

Your wrap design should be your own and have personal characteristics. Some agents put their face on their car and some don't depending on personal preference.

The main point here is to make your contact information prevalent. You want people to be able to easily remember your vehicle and how to contact you. Make your website and phone number large and easy to read. It's a great idea to hook up with a talented graphic designer for this process. Wrap companies usually employ graphic designers.

11. Putting It All Together

Now you're ready to begin putting it all together. The first components you can start with are your Tablet PC or laptop and a mounting device for it. When I built my first van, I had a mechanic friend of mine put in all of the hardware. This is of course up to you.

In terms of running cords, cleaner is always better. You don't want your clients stepping on or over cords or even having to look at them. When building my mobile office, I was very particular about the overall appearance. No one wants to ride in a car with wires strung everywhere. Think of your car as your office!

Please refer to page 15 for a complete list of everything you will need to build your mobile office. The diagrams on the following pages are suggestions for configuring your mobile office.

Configuration for a van

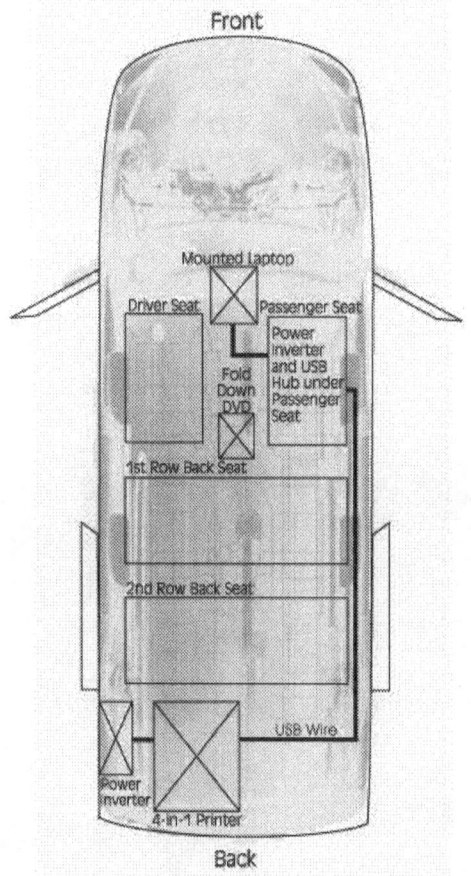

Front

Mounted Laptop

Driver Seat

Passenger Seat

Power inverter and USB Hub under Passenger Seat

Fold Down DVD

1st Row Back Seat

2nd Row Back Seat

USB Wire

Power inverter

4-in-1 Printer

Back

Configuration for a car

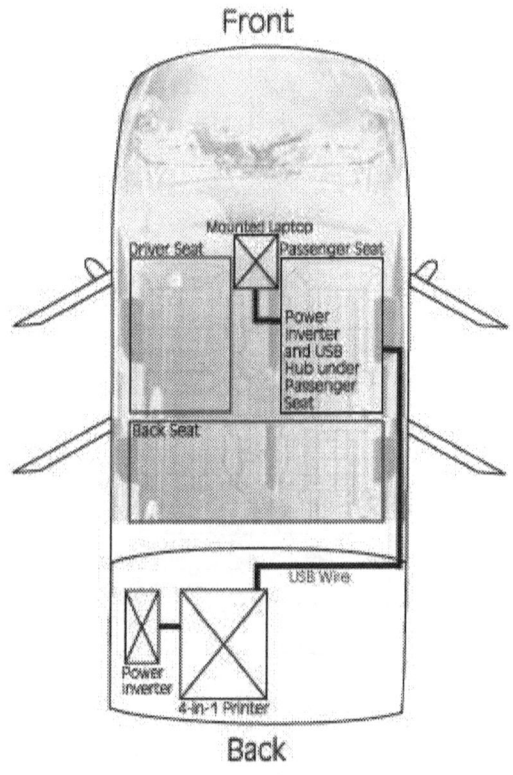

Front

Mounted Laptop

Driver Seat · Passenger Seat

Power inverter and USB Hub under Passenger Seat

Back Seat

USB Wire

Power inverter

4-in-1 Printer

Back

12. Product Suggestions

Again, these are in no way endorsements or promises of these products and services. They are based solely on personal experience. Because these companies are launching new products each month, I will not even attempt to give specific model numbers.

Product or Service	Suggestion(s)
Tablet PC or laptop	Toshiba, Dell, Gateway
Mounting device for Tablet PC or laptop	Jotto Desk®
Wireless Internet card	Verizon, Sprint, Cingular
Computer cords	Radio Shack
Power Inverters	Auto parts store or electronics store
E-fax service	www.efax.com or www.protus.com
Multi-functional printer	Lexmark, HP
GPS	Garmin, Magellan

13. Backing Up And Computer Health

Three very important words: Backup, Backup, Backup. Losing an entire hard drive will make a grown-up cry. I've seen it. Backing up is particularly important when working in the mobile and paperless office environment. Here are some tips on backing up and general computer health.

Backing Up Your Data

There are a variety of ways to keep a copy of your data outside of your laptop or Tablet PC's hard drive. I have found the easiest way is backing up to an external hard drive.

You can purchase an external hard drive at any computer or electronics store. A 30GB external hard drive will provide sufficient room unless you have a collection of particularly large files.

Keeping all of your files under "My Documents" will make it easy to backup. Just plug in your external hard drive, copy "My Documents" and paste into your external hard drive.

You can also backup through a Windows feature called "Backup." Just go to "Start, Programs, Accessories, System Tools, and then to Backup." You can then either backup everything or just the files that have changed since your last backup. Be sure you are backing up to your external hard drive. Some people prefer to backup on CD's. This is a more time consuming process, and is done simply by copying the "My Documents" folder to a CD using your computer's CD burner.

System Restore

A "restore point" will restore your computer to a specific point should you get a corrupted hard drive or virus. Every month or so, I make what is called a "Restore Point." This way if my hard drive gets corrupted I have a restore point I can easily reset my system to. The beauty of this feature is that it will also backup any programs you have installed, not just the files you have made in My Documents. To make a restore point, just go to "programs, accessories, system tools, and system restore."

Disk Defragmenter/Disk Cleanup

These programs are necessary to keep your hard drive running smoothly. You can schedule these programs to run automatically at any time, as long as your computer is turned on. To do this, go to "Start, Programs, Accessories, and System Tools.

Anti-Virus Protection

It goes without saying that you should always be running an anti-virus program. Popular programs are McAfee® and Norton AntiVirus™.

Spyware Protection

Spyware today has become more of a threat to your computer than viruses.

Popular anti-spyware programs are Spybot and Adware®. The personal versions are free, and there is a fee for a commercial use license.

14. Bonus Chapter Going Paperless

Save a tree and go paperless!

With the rapid advances in technology, real estate transactions are going paperless. Imagine never having large manila folders to carry around anymore. Going paperless is easy with mobile Internet and e-faxing technology. This will streamline the process and save you a lot of time by always having every document with you at all times.

Notice - Electronic signatures and facsimile documents may not be legally binding in all states. You might need original signatures. Check with your local real estate board.

When it comes to doing paperless transactions, I cannot stress enough that practice makes perfect. Practice all of these techniques before using them with a client. Make up sample contracts and go through a mock signing at home before going paperless in the field.

The Tablet PC plays an integral role in going paperless. Why? Because Tablet PC's offer the unique ability to sign directly on the screen. *Offers for property can now be filled out and signed on your Tablet PC and e-faxed immediately while in the field!*

The following is a detailed guide of how I cut down on paper and printing as much as possible. As I said earlier, ink is expensive!

There are many variations and new services coming out constantly. The following is a suggestion based on my experience in the field.

What you will need to go paperless

1. A Tablet PC running Windows XP or Windows 2000

2. Microsoft Office Image Document Writer or Windows Journal Writer (this should have already come preloaded on your Tablet PC)

3. A program called Cute PDF writer (see instructions below)

4. Zip Form® software (see below)

5. An established e-fax account (see previous chapter)

6. An email address

Zip Form®

Zip Form® may not be available in all states. Zip Form® is a program available to REALTORS® that already contains the documents you need to conduct real estate business in electronic format. Go to their website or check with your local association to get your licensed copy.

Zip Form® makes life simple by already having all of the relevant documents in electronic format. All you do is fill in the blanks on your computer. Your forms will look more professional when typed, and can save you a lot of time. Zip Form® requires a subscription. My local real estate board includes the subscription with our association fees. Check with your local board or go to www.ZipForm.com for product information.

Cute PDF Writer

Formally known as Cute PDF Printer, Cute PDF Writer is a free program that will convert what is shown on any Windows screen into a PDF document. Why would you want to do this? Because a PDF document cannot be altered once saved, and they are a standard format for e-faxing. To get a free copy of Cute PDF Writer, go to www.cutepdf.com

If you do not have Zip Form® installed, you may scan in your hand-signed documents using your multi-functional printer.

Microsoft Office Image Document Writer or Windows Journal Writer

This program should already be pre-loaded on your Tablet PC. To check to see if it is loaded on your computer, click on "File" and then "Print…" in any program such as Word. You should see Image Writer or Journal Writer as an option to print.

How It Works

That may sound like a lot, but it is only a total of 3 programs. Remember, practice, practice, practice.

1. Open Zip Form® and fill in all the necessary fields.

2. Once you have filled in all of the available fields, go to "File" at the top of the screen, then hit "Print…" Instead of choosing your default printer, *print the document to your Microsoft Office Image Document Writer*. It will ask you where you would like to save it. You can either save it on the desktop or in your client's file folder that you have created.

3. Find where you have saved the document and double click the icon to open in Image Document Writer.

4. Now, *swing* your Tablet PC screen around and lay it flat to go into "tablet mode."

5. The client(s) then initials and signs each document directly on the screen using the stylus as they normally would on paper.

6. After all signatures have been done, flip the screen back up to regular mode. Now hit "File" and "Print..." This time *print to Cute PDF Writer.* This should be in your list of print options once you have successfully installed this program. Again it will ask you where you would like to save it. Choose either the desktop or your client's folder.

7. Your signed document is now saved in PDF format! From here you can either e-fax the document(s), print, or email them as an attachment, right from your mobile office!

15. Conclusion

We are all looking for more time. Modern mobile computing makes this possible. Once you have built your mobile office, information will be available in a split second while in the field. Believe me; this can make a difference in getting a sale. Clients want information now!

The mobile office is inherently a buyer's tool. While you can apply these technologies to sellers as well, the simple act of being in the field with a buyer necessitates having the technology with you.

If the going seems slow while building your first office on wheels, don't be discouraged. The techniques in this book took years to collect and implement. Hopefully the instructions in this book will help speed up that process.

There are many resources for you as a REALTOR®. The National Association of REALTORS® has a technology section on their website, you can become e-Pro certified, or take classes through a local board or community college. Good luck!

16. Mobile Office Glossary

Air Card- Wireless PC card which will allow your computer to login to the Internet wherever there is coverage. (check with your local provider)

Blue Tooth- Provides a way to connect and exchange information between devices such as printers, laptops, Tablet PC's, or digital cameras via a secure radio frequency

E-Fax- Technology which will enable you to send and receive faxes directly on your computer. Inbound faxes come in as .PDF (you will need Adobe Reader® to view them) Outbound faxing is done through sending an email with the attached document(s) you want to fax

External Hard Drive- A hard drive which connects to your computer via a USB cord in order to backup your data

GPS- Global Positioning Satellite which enables you to pinpoint your location and will give directions to your next stop

Hard Drive- The place where your computer stores your documents and programs internally. Looks like a needle on a record player.

Hot Spot- Locations such as coffee shops or airports that offer a wireless connection to the Internet (sometimes for a fee) Hotspots are not necessary when using an air card

HUB- A device which allows many USB devices to be connected to a single USB port on your computer

Multifunctional Printer- A printer which also has copy, scan and fax capability

PDA- Personal digital assistant such as a Palm® Pilot

Power Inverter- A device which will convert the DC power in your car to the AC power your computer and printer need

Stylus- A pen type device used for writing directly on a Tablet PC or PDA screen

Tablet PC- Similar to a laptop, a Tablet PC has the unique capability to flip the screen down into "tablet" mode and write directly on the screen

USB Cable- A standard cable which connects peripheral devices to your computer such as a printer or keyboard. Some newer cables may be "USB2"

Ultra Portable PC- A computer offering full Windows functionality. Smaller than a traditional laptop or Tablet PC

Wrapping Your Vehicle- Putting high impact graphics on your car with your logo & contact information. Literally a moving billboard

Zip Form®- A program which provides real estate documents in electronic format